STRUCTURAL ANALYSIS AND DESIGN

Principles and Applications

SREEKUMAR V T

PREFACE

Welcome to 'Structural Analysis and Design: Principles and Applications'. This book aims to provide a comprehensive and practical guide to the fundamental principles and methodologies of structural engineering. Whether you are a student, a practicing engineer, or someone interested in the field of structural analysis and design, this book is designed to equip you with the knowledge and skills necessary to tackle real-world structural engineering challenges.

Structural engineering plays a crucial role in shaping our built environment. It is the discipline that enables us to design and construct safe, efficient, and aesthetically pleasing structures that can withstand the forces and loads they encounter. From towering skyscrapers to bridges, stadiums, and residential buildings, structures surround us and impact our lives every day. Understanding the principles of structural analysis and design is essential for creating structures that not only meet the functional requirements but also ensure the safety and well-being of the people who use them.

This book is structured to provide a logical progression of topics, starting with an introduction to structural engineering and the basic principles underlying structural analysis. We then delve into the different types of loads and load combinations that structures must be designed to withstand. The book covers the properties of structural materials, including concrete, steel, timber, and composites, and their influence on structural design.

The chapters on analysis of determinate and indeterminate

structures provide a comprehensive overview of the methods and techniques used to analyze structures under various loading conditions. We explore both classical and modern approaches, emphasizing the importance of understanding structural behavior and the underlying mathematical concepts.

In the subsequent chapters, we shift our focus to the design aspects of structural engineering. We discuss the design of beams, columns, tension members, and connections, providing practical guidelines and design methodologies based on industry codes and standards. The design of reinforced concrete structures and steel structures is covered in separate chapters, addressing the specific considerations and detailing requirements for each material.

Additionally, this book explores topics such as structural modeling, computer-aided analysis, structural stability, buckling analysis, composite structures, and seismic analysis. These topics are essential for understanding the complexities of modern structural engineering and the challenges associated with designing structures that can withstand dynamic loads and natural hazards.

Throughout the book, we strive to strike a balance between theoretical concepts and practical applications. We believe that a solid understanding of the underlying principles is crucial for effective structural analysis and design. However, we also recognize the importance of real-world examples, case studies, and practical insights that bridge the gap between theory and practice.

We have included numerous illustrations, diagrams, and solved examples to facilitate the learning process and enhance comprehension. The book also provides references to relevant codes, standards, and additional resources for readers interested in further exploration of specific topics.

It is important to note that this book serves as a comprehensive introduction to structural analysis and design, but it is not intended to replace engineering education or experience. It is meant to be a valuable resource that complements academic coursework, professional development, and self-study, providing readers with a solid foundation in the principles and applications of structural engineering.

We would like to express our gratitude to the numerous individuals, educators, and professionals who have contributed to the development of this book. Their expertise, insights, and dedication to the field of structural engineering have greatly influenced the content and structure of this work.

We hope that 'Structural Analysis and Design: Principles and Applications' proves to be a valuable resource in your journey to understand and apply the principles of structural engineering. May it inspire your curiosity, deepen your knowledge, and empower you to design structures that stand the test of time.

Happy reading and may you embark on a fulfilling and successful career in the fascinating world of structural engineering!

SREEKUMAR V T

CONTENTS

1. Introduction to Structural Engineering
2. Loads and Load Combinations
3. Properties of Structural Materials
4. Analysis of Determinate Structures
5. Analysis of Indeterminate Structures
6. Structural Modeling and Computer-Aided Analysis
7. Structural Stability and Buckling Analysis
8. Design of Beams and Flexural Members
9. Design of Columns and Compression Members
10. Design of Tension Members and Connections
11. Design of Reinforced Concrete Structures
12. Design of Steel Structures
13. Design of Composite Structures
14. Structural Dynamics and Earthquake Analysis
15. Introduction to Structural Design Codes and Standards

1. INTRODUCTION TO STRUCTURAL ENGINEERING

Structural engineering is a field of engineering that deals with the design, analysis, and construction of structures that can withstand the forces and loads they may encounter. It plays a crucial role in ensuring the safety, stability, and functionality of buildings, bridges, dams, towers, and other infrastructure.

The primary objective of structural engineering is to create structures that can withstand the various loads they may experience throughout their lifespan. These loads can include gravity loads (such as the weight of the structure itself and any occupants or equipment), environmental loads (such as wind, rain, snow, and earthquakes), and imposed loads (such as the weight of vehicles on a bridge or machinery in a building).

To achieve this objective, structural engineers utilize a combination of scientific principles, mathematical calculations, and engineering techniques. They consider the properties of construction materials, the behavior of structures under different loads, and the application of various design codes and standards.

Structural analysis is a fundamental aspect of structural engineering. It involves determining the internal forces and deformations in a structure due to the applied loads. By understanding these forces and deformations, engineers can assess the structural integrity and performance of a design. Analytical methods, such as the finite element method, are often employed to model and simulate complex structures.

Once the structural analysis is complete, the design phase begins. Structural engineers use their expertise to develop a safe and efficient design that can withstand the anticipated loads. They select appropriate materials, determine the size and configuration of structural members, and specify the connections between them. The design process also involves considering factors like serviceability, constructability, and sustainability.

Structural engineers work with a wide range of materials, including concrete, steel, timber, and composite materials. Each material has its own unique properties and behavior under different loads, which must be taken into account during the design process. For example, steel is known for its high strength and ductility, making it suitable for structures that require flexibility and resistance to dynamic forces. On the other hand, concrete offers durability and fire resistance, making it ideal for buildings and infrastructure projects.

The role of a structural engineer goes beyond just designing structures. They also play a crucial role in construction management and ensuring the integrity of the structure during the construction phase. Structural engineers collaborate with architects, contractors, and other professionals to oversee the implementation of the design, conduct inspections, and address any unforeseen challenges that may arise.

In recent years, advancements in technology have greatly impacted the field of structural engineering. Computer-aided design (CAD) software and structural analysis tools have revolutionized the way engineers design and analyze structures. These tools enable engineers to create detailed 3D models, perform intricate simulations, and optimize designs for cost-effectiveness and efficiency. Furthermore, the use of Building Information Modeling (BIM) has improved collaboration and coordination among various stakeholders involved in a construction project.

Another important aspect of structural engineering is the consideration of sustainability and resilience. Structural engineers strive to design structures that are not only safe and functional but also environmentally friendly and energy-efficient. They incorporate sustainable design practices, such as using recycled materials, optimizing energy consumption, and integrating renewable energy systems.

In conclusion, structural engineering is a vital discipline that ensures the safety, stability, and functionality of structures. By applying scientific principles, mathematical calculations, and engineering techniques, structural engineers design and analyze structures to withstand various loads and forces. They play a crucial role in the construction process, collaborating with other professionals to ensure the successful implementation of the design. With advancements in technology and a growing focus on sustainability, structural engineering continues to evolve and shape the way we build the infrastructure of the future.

2. LOADS AND LOAD COMBINATIONS

In structural engineering, the design and analysis of structures heavily rely on understanding the loads and load combinations that a structure may experience throughout its lifespan. Loads are external forces or deformations that act on a structure and influence its behavior. These loads can come from various sources, including environmental factors, occupancy, equipment, and construction activities. Proper consideration and analysis of these loads are essential to ensure the safety, reliability, and functionality of the structure. This article provides an overview of loads and load combinations in structural engineering.

Types of Loads:

Structural loads can be broadly categorized into two types: dead loads and live loads. Dead loads, also known as permanent loads or static loads, are the self-weight of the structure, including the weight of the structural elements, finishes, and fixed equipment. Dead loads are constant and remain in place for the duration of the structure's life. Examples of dead loads include the weight of the walls, floors, roofs, and permanent fixtures.

On the other hand, live loads, also referred to as variable loads or imposed loads, are loads that are not permanently fixed and can change in magnitude or location. Live loads are transient in nature and result from occupancy, moving loads, or temporary construction activities. Examples of live loads include the weight of people, furniture, vehicles, storage materials, and equipment. Live loads are considered as they can significantly impact the structural response and should be carefully evaluated during the design process.

In addition to dead loads and live loads, structures are also subjected to environmental loads, which can vary based on the geographical location and local conditions. Environmental loads include loads due to wind, snow, rain, earthquakes, and temperature variations. These loads can have a significant influence on the structural design and require thorough analysis and consideration.

Load Combinations:

Load combinations involve combining different types of loads to assess the overall effect on the structure. In practice, structures are subjected to multiple loads simultaneously, and it is crucial to consider the worst-case scenario when designing for strength, stability, and serviceability. Load combinations are developed based on established design codes and standards, which provide guidelines and criteria for the selection and combination of loads.

Design codes, such as the International Building Code (IBC), Eurocode, and American Society of Civil Engineers (ASCE) standards, define specific load combinations that structural engineers must consider during the design process. These codes outline various load combinations based on the type of structure, occupancy, and importance level. Load combinations typically include combinations of dead loads, live loads, and environmental loads, along with potential load factors or load reduction factors.

The load factors or load reduction factors applied to each load type in a combination account for uncertainties in load estimation and variations in load magnitude. These factors provide a safety margin to ensure that the structure can safely withstand the loads without excessive deformation or failure. The load factors are determined based on the probability of the loads occurring simultaneously or at their maximum expected values.

Load combinations are developed by considering different load cases and applying appropriate load factors. For example, a load combination for a residential building may consist of dead loads, live loads, wind loads, and snow loads. The load factors associated with each load type are selected based on the level of reliability required for the structure and the consequences of failure.

The process of load combination is iterative, as structural engineers analyze the effects of different load combinations on the structural elements and evaluate their capacity to resist the loads. This analysis involves determining the internal forces, stresses, and deformations in the structural members under the combined loads. The results of this analysis are then compared to the allowable limits defined by the design codes to ensure structural integrity and safety.

It is worth mentioning that load combinations can vary depending on the specific design code and local regulations. Therefore, structural engineers must stay updated with the latest versions of the applicable codes and standards to ensure compliance and best practices in load combination design.

Conclusion:

Loads and load combinations are fundamental aspects of structural engineering that influence the design and analysis of structures. By understanding and properly considering the different types of loads, including dead loads, live loads, and environmental loads, structural engineers can develop safe and reliable designs that can withstand the anticipated forces and deformations. Load combinations play a crucial role in assessing the worst-case scenario and accounting for uncertainties in load estimation. Through meticulous analysis and adherence to design codes, engineers ensure that structures are built to withstand a wide range of loads and provide the necessary safety and functionality for their intended purposes.

3. PROPERTIES OF STRUCTURAL MATERIALS

In structural engineering, the selection of appropriate materials is a critical aspect of designing and constructing safe, efficient, and durable structures. Structural materials possess specific properties that determine their behavior under various loads and conditions. Understanding the properties of structural materials is essential for structural engineers to make informed decisions and optimize the design and performance of structures. This article provides an overview of the properties of commonly used structural materials, including concrete, steel, timber, and composite materials.

Concrete:

Concrete is a versatile and widely used construction material in structural engineering. It consists of a mixture of cement, aggregates (such as sand and gravel), water, and often additional additives. Concrete exhibits several important properties that make it suitable for structural applications:

1. Compressive Strength: Concrete is known for its excellent compressive strength, which refers to its ability to resist forces that tend to squeeze or compress it. The compressive strength of concrete depends on the proportions of its constituents, the curing process, and other factors. Designers must specify the required compressive strength based on the structural requirements.

2. Tensile Strength: While concrete has relatively low tensile strength, it can still resist small tensile forces. However, to enhance its tensile strength and prevent cracking, reinforcement in the form of steel bars or fibers is commonly used in concrete structures.
3. Durability: Concrete has good durability and can withstand exposure to various environmental conditions, including moisture, chemicals, and temperature changes. Proper design, construction techniques, and protective measures are employed to ensure long-term durability.
4. Workability: Concrete can be molded and poured into different shapes and forms during construction, allowing for flexibility in design. The workability of concrete depends on factors such as water-cement ratio, aggregate size, and the use of admixtures.

Steel:

Steel is another widely used structural material known for its high strength and versatility. It is primarily an alloy of iron and carbon, with small amounts of other elements added to enhance specific properties. The key properties of steel include:

1. Strength: Steel has excellent strength properties, including high tensile and compressive strength. This makes it suitable for withstanding heavy loads and forces in structural applications.
2. Ductility: Steel exhibits high ductility, which means it can deform significantly before failure. This property allows steel structures to absorb energy and withstand dynamic loads, such as those caused by earthquakes or wind.
3. Elasticity: Steel has a high modulus of elasticity, which means it can deform elastically under loads and return to its original shape once the load is removed. This property helps maintain the structural integrity of steel

elements.
4. **Weldability:** Steel is highly weldable, allowing for efficient and cost-effective fabrication of steel structures. Welding provides strong connections between steel members, contributing to the overall strength of the structure.
5. **Corrosion Resistance:** Although steel is susceptible to corrosion, various protective measures such as coatings, galvanization, and stainless steel alloys can significantly enhance its corrosion resistance.

Timber:

Timber, or wood, has been used as a structural material for centuries. It offers unique properties and advantages for certain applications. The properties of timber include:

1. **Strength:** Timber has good strength properties, with different types of wood exhibiting varying levels of strength. Strength is influenced by factors such as species, density, and grain orientation.
2. **Lightweight:** Timber is a lightweight material, which makes it easy to handle and transport. This property can reduce construction costs and simplify the construction process.
3. **Thermal Insulation:** Wood has excellent thermal insulation properties, providing natural insulation against heat transfer. This can contribute to energy efficiency in buildings.
4. **Environmental Sustainability:** Timber is a renewable resource, and its use in construction can contribute to sustainable practices and reduce the carbon footprint of structures.

Composite Materials:

Composite materials are engineered materials that combine two or more constituent materials to create a material with enhanced properties. Common examples of composite materials used in structural engineering include fiber-reinforced polymers (FRPs) and reinforced concrete. The properties of composite materials depend on the combination and arrangement of the constituent materials, resulting in specific advantages such as:

1. High Strength-to-Weight Ratio: Composite materials often have exceptional strength-to-weight ratios, making them lightweight yet strong. This property allows for the design of lightweight structures that can withstand significant loads.
2. Corrosion Resistance: Composite materials, such as FRPs, are typically resistant to corrosion, making them suitable for applications in aggressive environments or structures exposed to corrosive elements.
3. Design Flexibility: Composite materials offer design flexibility, as they can be molded and shaped into complex forms. This enables the creation of innovative and aesthetically pleasing structures.
4. Fatigue Resistance: Many composite materials exhibit excellent fatigue resistance, making them suitable for structures subjected to cyclic loads or dynamic forces.

Conclusion:

Understanding the properties of structural materials is crucial for structural engineers in the design and construction of safe and efficient structures. Concrete, steel, timber, and composite materials each possess unique properties that influence their behavior under various loads and environmental conditions. By considering these properties, structural engineers can select the most appropriate material for a given application, optimize the design, and ensure the structural integrity, durability, and functionality of the constructed structure. Proper material selection, coupled with accurate analysis and adherence to design codes and standards, contributes to the overall success and longevity of structural projects.

4. ANALYSIS OF DETERMINATE STRUCTURES

In the field of structural engineering, the analysis of determinate structures is a fundamental process that enables engineers to understand the behavior and response of various types of structures under different loads. Determinate structures are those in which the reactions, internal forces, and deformations can be completely determined using the principles of statics. The analysis of determinate structures involves applying mathematical and engineering principles to evaluate the stability, strength, and overall performance of a structure. This article provides an overview of the analysis of determinate structures, including the methods and techniques used in the process.

1. Overview of Determinate Structures:

Determinate structures are those that can be analyzed using basic statics principles without the need for complex mathematical techniques. These structures have a predictable response to applied loads, and their internal forces and reactions can be determined through static equilibrium equations. Examples of determinate structures include simply supported beams, trusses, and frames with a limited number of supports and members.

2. Types of Loads on Determinate Structures:

Determining the behavior of a structure requires considering the various types of loads it may experience. These loads can include dead loads, live loads, wind loads, snow loads, and seismic loads, among others. Dead loads are the permanent weight of the structure itself and any fixed components, while live loads are transient loads that vary with occupancy or usage. Wind loads are caused by wind forces acting on the structure, and snow loads occur due to the weight of accumulated snow. Seismic loads result from earthquakes and induce dynamic forces on the structure. Analyzing the effects of these loads helps in understanding the internal forces and deformations in the structure.

3. Structural Analysis Methods for Determinate Structures:

a. Method of Joints: The method of joints is commonly used for analyzing trusses, which are skeletal structures composed of straight members connected by joints. In this method, the equilibrium of forces at each joint is considered to determine the internal forces in the truss members. By sequentially analyzing the joints, the complete internal force distribution in the truss can be obtained.

b. Method of Sections: The method of sections is another widely used technique for analyzing determinate structures, including beams and frames. This method involves cutting through a section of the structure and analyzing the equilibrium of forces and moments acting on the section. By examining the forces and moments on the section, the internal forces and reactions at specific locations can be determined.

c. Slope-Deflection Method: The slope-deflection method is employed for analyzing determinate beams and frames with more complex conditions, such as continuous beams or structures with multiple supports and members. This method involves considering the deflections and rotations at the supports and applying compatibility equations to determine the internal forces and reactions in the structure.

d. Moment Distribution Method: The moment distribution method is used for analyzing continuous beams and frames. It involves distributing moments at the supports and iteratively redistributing them until equilibrium is achieved. The method considers the stiffness of the structure and the rotational behavior at the supports to calculate the internal forces and reactions.

4. Structural Analysis Software:

With advancements in technology, structural analysis software has become an integral tool for analyzing determinate structures. These software applications employ numerical methods and algorithms to perform complex calculations and provide accurate results. Structural analysis software allows engineers to input the geometric and material properties of the structure, specify the loads, and obtain detailed information about the internal forces, reactions, and deformations. Such software simplifies the analysis process and enables engineers to explore different design scenarios and optimize the structure's performance.

5. Limitations of Determinate Structures:

While determinate structures have advantages in terms of simplicity and ease of analysis, they also have certain limitations. Determinate structures may not accurately represent real-world structures that are subjected to complex loading conditions or possess non-linear material properties. Additionally, determinate structures do not account for structural redundancy, where the failure of one component may not cause complete collapse due to alternative load paths. For more complex structures, indeterminate analysis methods, such as the matrix stiffness method or finite element method, are required to accurately capture the structural behavior.

Conclusion:

The analysis of determinate structures is a fundamental process in structural engineering that allows engineers to understand the behavior and response of structures under different loads. By considering the types of loads and employing various analysis methods, such as the method of joints, method of sections, slope-deflection method, and moment distribution method, engineers can determine the internal forces and reactions in determinate structures. The use of structural analysis software further simplifies the analysis process and provides accurate results. While determinate structures have limitations, their analysis provides valuable insights into the stability, strength, and overall performance of structures, contributing to the safe and efficient design of various engineering projects.

5. ANALYSIS OF INDETERMINATE STRUCTURES

In the field of structural engineering, the analysis of indeterminate structures plays a crucial role in designing complex and efficient structures. Unlike determinate structures, indeterminate structures are those that cannot be fully analyzed using basic statics principles alone. These structures possess more unknowns than available equilibrium equations, requiring the application of advanced analysis techniques and methods. The analysis of indeterminate structures allows engineers to accurately determine internal forces, deformations, and reactions, leading to safer and more optimized designs. This article provides an overview of the analysis of indeterminate structures, including the methods and techniques used in the process.

1.Understanding Indeterminacy:

Indeterminacy in structures refers to the condition where the number of unknowns exceeds the number of equilibrium equations available for analysis. In other words, the internal forces and reactions in indeterminate structures cannot be determined by static equilibrium alone. This occurs due to the presence of redundant members or redundant supports, which offer alternative load paths and introduce additional unknowns.

2. Types of Indeterminate Structures:

Indeterminate structures can be broadly classified into two types: statically indeterminate and kinematically indeterminate structures.

a. Statically Indeterminate Structures: Statically indeterminate structures have redundant members or supports that result in more unknowns than available equilibrium equations. These structures require additional analysis techniques beyond statics principles to determine the internal forces and reactions accurately. Examples of statically indeterminate structures include continuous beams, frames with multiple supports, and structures with redundant members.

b. Kinematically Indeterminate Structures: Kinematically indeterminate structures are those that have additional constraints or redundancies in terms of their deformations or rotations. These structures cannot be fully analyzed using only statics principles but require consideration of compatibility and deformation equations. Examples of kinematically indeterminate structures include arches, cable structures, and some types of trusses.

3. Methods for Analyzing Indeterminate Structures:

To analyze indeterminate structures, engineers employ advanced analysis methods that go beyond basic statics principles. These methods take into account the additional unknowns and consider the structural behavior, including deformations and rotations.

a. Flexibility Method: The flexibility method, also known as the force method, is commonly used to analyze indeterminate structures. This method involves assuming a known displacement or rotation at one or more points in the structure and then determining the corresponding forces and moments throughout the structure. By solving a system of equations based on equilibrium and compatibility conditions, the internal forces and reactions can be determined.

b. Stiffness Method: The stiffness method, also known as the displacement method or the matrix method, is another widely used technique for analyzing indeterminate structures.

This method involves representing the structure as a series of interconnected elements and assembling stiffness matrices to describe the structural behavior. By applying compatibility equations and solving a system of equations, the displacements, internal forces, and reactions in the structure can be determined.

c. Finite Element Method (FEM): The finite element method is a powerful numerical technique used to analyze complex structures, including both indeterminate and determinate structures. FEM divides the structure into small elements, where each element is modeled mathematically using interpolation functions. By assembling these elements and solving the resulting system of equations, the displacements, internal forces, and reactions in the structure can be determined accurately.

4. Structural Analysis Software:

Given the complexity of analyzing indeterminate structures, structural analysis software has become an invaluable tool for structural engineers. These software applications employ advanced algorithms and numerical methods to solve the complex equations involved in the analysis of indeterminate structures. By inputting the geometric and material properties of the structure, specifying the loads, and defining the support conditions, engineers can obtain detailed information about the internal forces, deformations, and reactions. Structural analysis software simplifies the analysis process, enhances accuracy, and enables engineers to explore various design scenarios and optimize the structure's performance.

5. Importance of Indeterminate Analysis:

Analyzing indeterminate structures is of significant importance in structural engineering. Indeterminate analysis allows for a more accurate understanding of the internal forces, deformations, and reactions in complex structures, leading to safer designs that can withstand the anticipated loads and conditions. By considering the redistributions of forces and the behavior of redundant members or supports, engineers can optimize the design, reduce material usage, and enhance the overall efficiency of the structure. Indeterminate analysis also enables engineers to assess the effects of temperature changes, settlement, and other factors that can introduce additional complexities into the structural behavior.

Conclusion:

The analysis of indeterminate structures is a critical process in structural engineering that enables engineers to understand the behavior and response of complex structures. By employing advanced analysis methods such as the flexibility method, stiffness method, and finite element method, engineers can accurately determine the internal forces, deformations, and reactions in indeterminate structures. The use of structural analysis software further enhances the analysis process and provides detailed information for design optimization. Analyzing indeterminate structures is crucial for ensuring the safety, reliability, and efficiency of structures in various engineering projects. It allows for the consideration of complex load distributions, the behavior of redundant members, and the effects of deformations, resulting in designs that meet performance requirements and withstand the anticipated loads and conditions.

6. STRUCTURAL MODELING AND COMPUTER-AIDED ANALYSIS

In the field of structural engineering, the process of structural modeling and computer-aided analysis plays a vital role in designing safe, efficient, and cost-effective structures. Structural modeling involves creating a virtual representation of a physical structure using specialized software, while computer-aided analysis refers to the use of computational tools to evaluate the behavior and response of the structure under various loading conditions. This article provides an overview of structural modeling and computer-aided analysis, discussing their importance, methods, and benefits in the field of structural engineering.

1. Importance of Structural Modeling and Analysis:

Structural modeling and computer-aided analysis are essential steps in the design process of structures. They allow engineers to accurately predict and evaluate the structural behavior, including the internal forces, deformations, and stresses that the structure will experience. By simulating the response of the structure to different loading conditions, engineers can identify potential design flaws, optimize the structure's performance, and ensure its safety and durability. Structural modeling and analysis help in making informed design decisions, reducing costs, and avoiding catastrophic failures during construction or the structure's service life.

2. Types of Structural Modeling:

Structural modeling involves creating a digital representation of a physical structure using specialized software. There are various types of structural modeling techniques employed in practice:

a. 2D Modeling: In 2D modeling, the structure is represented in two dimensions, typically in plan or elevation views. This type of modeling is suitable for simpler structures, such as beams and frames, where the behavior can be adequately captured in a 2D plane.

b. 3D Modeling: 3D modeling involves creating a three-dimensional representation of the structure. It provides a more realistic and comprehensive view, considering all three dimensions. 3D modeling is commonly used for complex structures, such as buildings and bridges, where the behavior in all directions is significant.

c. Finite Element Modeling: Finite element modeling (FEM) is a widely used technique in structural engineering. It involves dividing the structure into small elements and analyzing each element individually. FEM allows for detailed modeling of complex structures and the consideration of material nonlinearities and geometric complexities.

d. Building Information Modeling: Building Information Modeling (BIM) is a comprehensive modeling approach that incorporates geometric, spatial, and non-geometric information about a structure. BIM includes not only the physical aspects of the structure but also attributes such as material properties, cost estimation, and scheduling. BIM facilitates collaboration between different stakeholders in the design, construction, and operation phases of a structure.

3. Computer-Aided Analysis Methods:

Computer-aided analysis involves using specialized software to simulate and evaluate the behavior of structures under various loading conditions. Different methods are used in computer-aided analysis, depending on the complexity of the structure and the desired level of accuracy:

a. Static Analysis: Static analysis involves evaluating the equilibrium of forces and moments in a structure under static loading conditions. It helps determine the internal forces, deformations, and stresses in the structure. Static analysis is suitable for structures subjected to constant or slowly varying loads.

b. Dynamic Analysis: Dynamic analysis considers the response of structures to dynamic or time-varying loads, such as earthquakes, wind gusts, or machinery vibrations. It accounts for the inertia and damping effects, providing valuable insights into the dynamic behavior and response of the structure.

c. Modal Analysis: Modal analysis focuses on determining the natural frequencies and mode shapes of a structure. It helps identify potential vibration issues and assesses the dynamic characteristics of the structure.

d. Nonlinear Analysis: Nonlinear analysis is used when the behavior of the structure deviates from linear assumptions. It considers nonlinear material properties, geometric nonlinearities, and large deformations that occur in certain structures or under specific loading conditions.

4. Benefits of Structural Modeling and Computer-Aided Analysis:

The adoption of structural modeling and computer-aided analysis brings several benefits to the field of structural engineering:

a. Improved Design Efficiency: Structural modeling and analysis tools enable engineers to explore multiple design alternatives and evaluate their performance quickly. This leads to more efficient and optimized designs, reducing material usage and construction costs.

b. Enhanced Safety and Reliability: By simulating the behavior of structures under different loading conditions, engineers can identify potential structural weaknesses and failure modes. This allows for the implementation of appropriate design modifications, ensuring the safety and reliability of the structure.

c. Cost and Time Savings: Structural modeling and analysis help optimize the design process, reducing the need for physical prototypes and costly modifications during construction. By identifying design issues early on, significant cost and time savings can be achieved.

d. Increased Design Flexibility: With computer-aided analysis, engineers can easily modify and iterate designs to meet specific project requirements. The ability to quickly assess the effects of design changes facilitates innovation and the exploration of alternative solutions.

e. Improved Communication and Collaboration: Structural modeling and analysis tools enable effective communication and collaboration among different stakeholders involved in a project. The visual representation of the structure and the analysis results facilitate better understanding and decision-making.

5. Limitations and Challenges:

While structural modeling and computer-aided analysis offer numerous advantages, there are some limitations and challenges to consider:

a. Accuracy and Assumptions: The accuracy of analysis results depends on the quality of the input data and the assumptions made during the modeling and analysis process. Incorrect assumptions or inadequate data can lead to inaccurate predictions of the structure's behavior.

b. Complexity of Analysis: Analyzing complex structures using advanced analysis methods such as FEM can be computationally intensive and time-consuming. The complexity of the analysis increases with the number of elements and degrees of freedom in the model.

c. Expertise and Training: Effective utilization of structural modeling and analysis tools requires expertise and training. Engineers need to have a solid understanding of structural behavior, analysis methods, and software capabilities to produce reliable and accurate results.

d. Software Limitations: Different software tools have their own limitations and capabilities. Engineers should be aware of these limitations and select appropriate software that suits the specific needs of the project.

Conclusion:

Structural modeling and computer-aided analysis have revolutionized the field of structural engineering. They enable engineers to accurately predict and evaluate the behavior of structures, leading to safer, more efficient, and cost-effective designs. Through various modeling techniques and analysis methods, engineers can simulate the response of structures to different loading conditions, identify potential design flaws, optimize performance, and ensure structural integrity. The adoption of structural modeling and computer-aided analysis brings significant benefits, including improved design efficiency, enhanced safety, cost and time savings, increased design flexibility, and improved collaboration. However, it is crucial to consider the limitations and challenges associated with these tools and to employ them with sound engineering judgment and expertise. By leveraging the power of structural modeling and computer-aided analysis, structural engineers can push the boundaries of design and create innovative structures that meet the demands of our ever-changing built environment.

7. STRUCTURAL STABILITY AND BUCKLING ANALYSIS

Structural stability is a critical consideration in the field of structural engineering. It refers to the ability of a structure to resist deformation or failure under various loading conditions. When a structure loses its stability, it undergoes a phenomenon known as buckling, which can lead to catastrophic failure. Understanding the concepts of structural stability and performing buckling analysis are essential for designing safe and reliable structures. This article provides an overview of structural stability and buckling analysis, discussing their significance, factors affecting stability, and methods employed in the analysis.

1. Importance of Structural Stability:

Structural stability is crucial for the safety and performance of structures. When a structure loses stability, it can experience sudden and unpredictable deformations or failure, posing significant risks to life and property. Therefore, ensuring the stability of structures is of paramount importance in structural engineering. By understanding the factors that affect stability and employing appropriate design and analysis techniques, engineers can create structures that can withstand the anticipated loads and remain stable throughout their service life.

2. Factors Affecting Structural Stability:

Several factors influence the stability of structures. Understanding these factors is essential for analyzing and designing stable structures:

a. Load Magnitude and Distribution: The magnitude and distribution of applied loads significantly impact the stability of structures. Excessive loads or uneven load distribution can cause local instabilities, leading to buckling or collapse.

b. Material Properties: The mechanical properties of the materials used in the structure, such as elasticity, yield strength, and stiffness, affect its stability. Inadequate material strength or stiffness can result in structural instability.

c. Geometric Imperfections: Imperfections in the geometry of the structure, such as initial crookedness, warping, or deviations from straightness, can reduce stability. These imperfections create local stress concentrations that may initiate buckling.

d. Support Conditions: The support conditions of a structure play a significant role in its stability. Inadequate or improper supports can cause instability by inducing excessive displacements or rotations.

e. Slenderness Ratio: The slenderness ratio, defined as the ratio of the length of a structural member to its cross-sectional dimensions, influences stability. Members with high slenderness ratios are more susceptible to buckling than those with low slenderness ratios.

3. Buckling and Its Types:

Buckling is a phenomenon that occurs when a slender structural member loses stability and undergoes large deformations or failure. Buckling can occur in various forms, depending on the type of structure and loading conditions:

a. Euler Buckling: Euler buckling is the most fundamental form of buckling and occurs in idealized slender columns subjected to axial compression. It is characterized by a sudden lateral deflection of the column.

b. Lateral-Torsional Buckling: Lateral-torsional buckling occurs in beams or beam-like members subjected to combined axial compression and bending moments. It involves lateral deflection and twisting of the member.

c. Plate Buckling: Plate buckling occurs in thin plates subjected to compressive loads. It leads to deformations such as wrinkling, oil-canning, or local buckling.

d. Global Buckling: Global buckling refers to the overall instability of an entire structure or a significant portion of it. It can occur in various types of structures and is typically caused by a combination of compression, bending, and shear forces.

4. Buckling Analysis Methods:

To assess the stability of structures and predict the occurrence of buckling, engineers employ various analysis methods:

a. Theoretical Analysis: Theoretical analysis involves using mathematical equations and analytical methods to determine critical loads and buckling modes. The Euler buckling equation and various stability criteria are used in theoretical analyses.

b. Experimental Testing: Experimental testing involves subjecting physical models or prototypes to controlled loading conditions to observe their behavior and identify the onset of instability. It can provide valuable data for validating theoretical predictions and calibrating design assumptions.

c. Finite Element Analysis (FEA): Finite element analysis is a numerical method widely used in structural engineering. It involves dividing the structure into smaller elements and solving the equations of equilibrium to determine the critical loads and buckling modes.

d. Stability Design Codes: Many design codes and standards provide guidelines and procedures for evaluating structural stability and performing buckling analysis. These codes specify safety factors, design methods, and criteria to ensure stable and safe structures.

5. Design Considerations for Structural Stability:

Designing for structural stability requires careful consideration of various factors:

a. Proper Member Sizing and Spacing: Selecting appropriate member sizes and spacing is crucial for ensuring stability. Adequate cross-sectional dimensions and suitable member arrangements help resist buckling and maintain stability.

b. Bracing and Support Systems: Incorporating bracing and support systems can enhance stability. Bracing elements such as diagonal members or shear walls help restrain lateral deflections and prevent buckling.

c. Material Selection: Choosing materials with appropriate strength and stiffness properties is important for structural stability. Ensuring that the materials can resist the anticipated loads without undergoing instability is critical.

d. Geometric Design Considerations: Paying attention to geometric imperfections and minimizing deviations from the idealized geometry helps maintain stability. Proper detailing and fabrication techniques can reduce the likelihood of buckling.

e. Load Path Optimization: Designing an efficient load path that redistributes loads and minimizes stress concentrations can improve stability. This involves carefully routing loads from the source to the supports.

Conclusion:

Structural stability and buckling analysis are essential considerations in the design and analysis of structures. The ability of a structure to resist buckling and remain stable under various loading conditions ensures its safety and reliability. By understanding the factors that affect stability, employing appropriate analysis methods, and following design considerations, engineers can create stable structures that can withstand the anticipated loads and deformations. Buckling analysis techniques, including theoretical analysis, experimental testing, finite element analysis, and compliance with design codes, enable engineers to assess the stability of structures and predict the onset of buckling. By prioritizing structural stability in the design process, engineers contribute to the safety, durability, and performance of structures in various engineering projects.

8. DESIGN OF BEAMS AND FLEXURAL MEMBERS

Beams and flexural members are fundamental components of structural systems. They are responsible for carrying loads and resisting bending moments, which are common in many structural applications. The design of beams and flexural members involves ensuring their capacity to safely withstand the applied loads and maintain structural integrity. This article provides an overview of the design principles and considerations involved in designing beams and flexural members, discussing key concepts, design methods, and important considerations.

1. Introduction to Beams and Flexural Members:

Beams and flexural members are structural elements that primarily resist bending loads. They are subjected to moments that cause curvature and induce internal stresses within the material. Beams are typically horizontal or inclined structural members that span between supports, while flexural members encompass a broader category of structural elements that include beams, slabs, and other components that undergo bending deformation.

2. Design Considerations:

The design of beams and flexural members involves several considerations to ensure their strength, serviceability, and overall performance:

a. Load Types: Beams and flexural members are designed to resist various types of loads, including dead loads, live loads, wind loads, and seismic loads. Understanding the nature and magnitude of these loads is crucial for proper design.

b. Material Selection: The choice of material for beams and flexural members depends on factors such as strength requirements, durability, cost, and availability. Common materials used include reinforced concrete, structural steel, timber, and composite materials.

c. Limit States: The design of beams and flexural members is based on meeting specific limit states, including ultimate limit state (ULS) and serviceability limit state (SLS). The ULS ensures that the structure can safely withstand the maximum expected loads, while the SLS ensures that the structure remains functional and meets serviceability requirements.

d. Design Codes and Standards: Designing beams and flexural members involves adhering to relevant design codes and standards. These codes provide guidelines, equations, and safety factors to ensure the structural elements meet specified requirements and performance criteria.

3. Design Methods:

There are various methods and approaches for designing beams and flexural members. The selection of the design method depends on factors such as the material used, loading conditions, and structural requirements:

a. Working Stress Design (WSD): The working stress design method involves limiting the stress levels in the material to a fraction of its yield strength. This method is commonly used for materials such as timber and reinforced concrete.

b. Load and Resistance Factor Design (LRFD): The LRFD method, also known as the Load and Resistance Factor Rating (LRFR) method, uses load factors and resistance factors to ensure a sufficient safety margin against failure. It is commonly employed for structural steel design.

c. Ultimate Strength Design (USD): The USD method is based on the ultimate strength capacity of the material and its resistance against failure. It is commonly used in the design of reinforced concrete structures.

4. Design Process for Beams and Flexural Members:

The design process for beams and flexural members involves several steps:

a. Determining Design Loads: The first step is to determine the design loads acting on the beam or flexural member. These include dead loads, live loads, and any other applicable loads specified by the design criteria.

b. Selecting Appropriate Material: Based on the design loads and structural requirements, an appropriate material is selected. Factors such as strength, durability, and cost are considered.

c. Analyzing the Structure: The next step is to analyze the structure to determine the internal forces, moments, and shear forces acting on the beam or flexural member. This can be done analytically or by using structural analysis software.

d. Sizing and Designing the Member: Using the determined internal forces, the beam or flexural member is sized and designed to resist bending moments, shear forces, and any other relevant forces. Design equations, formulas, and code provisions are utilized for this purpose.

e. Checking Deflection and Serviceability: In addition to strength considerations, deflection and serviceability requirements must be checked to ensure the member meets performance criteria. Excessive deflections can affect the functionality and aesthetics of the structure.

f. Detailing and Reinforcement: If the member is made of reinforced concrete or composite materials, appropriate reinforcement is detailed based on design requirements. This includes selecting bar sizes, spacing, and detailing reinforcement arrangements.

g. Construction Considerations: The final step involves considering construction constraints and detailing requirements. Adequate formwork, construction joints, and reinforcement placement details are essential for proper construction and performance.

5. Advanced Design Considerations:

In addition to the basic design considerations, several advanced design aspects can be incorporated for optimizing the performance and efficiency of beams and flexural members:

a. Composite Action: Composite beams, consisting of a steel section with a concrete slab, can provide enhanced strength, stiffness, and efficiency. Composite action between the steel and concrete components is utilized to improve load-carrying capacity.

b. Prestressing and Post-Tensioning: Pre- or post-tensioning techniques can be employed to introduce compressive stresses in the member, enhancing its strength and reducing the effects of tensile forces.

c. Structural Optimization: Structural optimization techniques, such as topology optimization or shape optimization, can be used to find the most efficient and material-saving configurations for beams and flexural members.

d. Sustainability Considerations: Designing for sustainability involves considering factors such as material recyclability, embodied carbon, and life cycle assessment to minimize the environmental impact of beams and flexural members.

Conclusion:

The design of beams and flexural members is a critical aspect of structural engineering. By considering factors such as load types, material selection, limit states, and design methods, engineers can create safe, efficient, and reliable structures. The design process involves determining the design loads, selecting appropriate materials, analyzing the structure, sizing and designing the member, checking for deflection and serviceability, detailing reinforcement if required, and considering construction constraints. Advanced design considerations, such as composite action, prestressing, structural optimization, and sustainability, can further enhance the performance and efficiency of beams and flexural members. By employing sound engineering principles, adhering to design codes and standards, and considering advanced design techniques, engineers can ensure the successful design of beams and flexural members in a wide range of structural applications.

9. DESIGN OF COLUMNS AND COMPRESSION MEMBERS

Columns and compression members play a crucial role in structural systems by carrying vertical loads and resisting compressive forces. They are essential components in buildings, bridges, and other structures where they support the weight of the structure above and transfer it to the foundation. The design of columns and compression members involves ensuring their capacity to safely resist compression and maintain structural stability. This article provides an overview of the design principles and considerations involved in designing columns and compression members, discussing key concepts, design methods, and important considerations.

1. Introduction to Columns and Compression Members:

Columns and compression members are structural elements designed to primarily resist axial compressive forces. They are subjected to compressive loads that tend to cause shortening of the member. Columns are vertical or inclined structural members that transfer the loads from the superstructure to the foundation, while compression members encompass a broader category of elements that include beams, columns, trusses, and other components that undergo compression.

2. Design Considerations:

The design of columns and compression members involves several considerations to ensure their strength, stability, and overall performance:

a. Load Types: Columns and compression members are designed to resist various types of loads, including dead loads, live loads, wind loads, and seismic loads. Understanding the nature and magnitude of these loads is crucial for proper design.

b. Material Selection: The choice of material for columns and compression members depends on factors such as strength requirements, durability, cost, and availability. Common materials used include reinforced concrete, structural steel, timber, and composite materials.

c. Limit States: The design of columns and compression members is based on meeting specific limit states, including ultimate limit state (ULS) and serviceability limit state (SLS). The ULS ensures that the structure can safely withstand the maximum expected loads, while the SLS ensures that the structure remains functional and meets serviceability requirements.

d. Design Codes and Standards: Designing columns and compression members involves adhering to relevant design codes and standards. These codes provide guidelines, equations, and safety factors to ensure the structural elements meet specified requirements and performance criteria.

3.Design Methods:

There are various methods and approaches for designing columns and compression members. The selection of the design method depends on factors such as the material used, loading conditions, and structural requirements:

a. Working Stress Design (WSD): The working stress design method involves limiting the stress levels in the material to a fraction of its yield strength. This method is commonly used for materials such as timber and reinforced concrete.

b. Load and Resistance Factor Design (LRFD): The LRFD method, also known as the Load and Resistance Factor Rating (LRFR) method, uses load factors and resistance factors to ensure a sufficient safety margin against failure. It is commonly employed for structural steel design.

c. Strength Design: The strength design method focuses on ensuring that the column or compression member can safely resist the applied compressive loads without failure. It is commonly used in the design of reinforced concrete structures.

4. Design Process for Columns and Compression Members:

The design process for columns and compression members involves several steps:

a. Determining Design Loads: The first step is to determine the design loads acting on the column or compression member. These include dead loads, live loads, and any other applicable loads specified by the design criteria.

b. Selecting Appropriate Material: Based on the design loads and structural requirements, an appropriate material is selected. Factors such as strength, durability, and cost are considered.

c. Analyzing the Structure: The next step is to analyze the structure to determine the internal forces and moments acting on the column or compression member. This can be done analytically or by using structural analysis software.

d. Sizing and Designing the Member: Using the determined internal forces, the column or compression member is sized and designed to resist the applied compressive loads. Design equations, formulas, and code provisions are utilized for this purpose.

e. Checking Stability: Ensuring stability is a critical aspect of column and compression member design. Stability checks are performed to ensure that the member can resist buckling and maintain structural integrity under the applied compressive loads.

f. Detailing and Reinforcement: If the member is made

of reinforced concrete or composite materials, appropriate reinforcement is detailed based on design requirements. This includes selecting bar sizes, spacing, and detailing reinforcement arrangements.

g. Construction Considerations: The final step involves considering construction constraints and detailing requirements. Adequate formwork, construction joints, and reinforcement placement details are essential for proper construction and performance.

5. Advanced Design Considerations:

In addition to the basic design considerations, several advanced design aspects can be incorporated for optimizing the performance and efficiency of columns and compression members:

a. Composite Columns: Composite columns, consisting of a steel section encased in concrete, can provide enhanced strength, stiffness, and efficiency. Composite action between the steel and concrete components is utilized to improve load-carrying capacity.

b. Bracing Systems: Incorporating appropriate bracing systems can enhance the stability and load-carrying capacity of columns and compression members. Bracing elements such as diagonal members, shear walls, or truss systems help resist lateral forces and prevent buckling.

c. Structural Optimization: Structural optimization techniques, such as topology optimization or shape optimization, can be used to find the most efficient and material-saving configurations for columns and compression members.

d. Sustainability Considerations: Designing for sustainability involves considering factors such as material recyclability, embodied carbon, and life cycle assessment to minimize the environmental impact of columns and compression members.

Conclusion:

The design of columns and compression members is a critical aspect of structural engineering. By considering factors such as load types, material selection, limit states, and design methods, engineers can create safe, efficient, and reliable structures. The design process involves determining the design loads, selecting appropriate materials, analyzing the structure, sizing and designing the member, checking for stability, detailing reinforcement if required, and considering construction constraints. Advanced design considerations, such as composite columns, bracing systems, structural optimization, and sustainability, can further enhance the performance and efficiency of columns and compression members. By employing sound engineering principles, adhering to design codes and standards, and considering advanced design techniques, engineers can ensure the successful design of columns and compression members in a wide range of structural applications.

10. DESIGN OF TENSION MEMBERS AND CONNECTIONS

Tension members and their connections play a vital role in structural engineering by carrying tensile forces and transmitting loads. They are extensively used in various applications, including bridges, buildings, and industrial structures. The design of tension members and their connections involves ensuring their capacity to safely resist tension and maintain structural stability. This article provides an overview of the design principles and considerations involved in designing tension members and connections, discussing key concepts, design methods, and important considerations.

1. Introduction to Tension Members and Connections:

Tension members are structural elements designed to primarily resist axial tensile forces. They are subjected to pulling forces that tend to elongate the member. Tension members are typically cables, rods, or bars that carry the applied loads and transmit them to the supporting elements or connections. Connections are the junction points where tension members are attached or joined to other structural components.

2. Design Considerations:

The design of tension members and connections involves several considerations to ensure their strength, stability, and overall performance:

a. Load Types: Tension members are designed to resist various types of loads, including dead loads, live loads, wind loads, and seismic loads. Understanding the nature and magnitude of these loads is crucial for proper design.

b. Material Selection: The choice of material for tension members depends on factors such as strength requirements, corrosion resistance, cost, and availability. Common materials used include structural steel, high-strength alloys, and cables made of materials such as steel or carbon fiber.

c. Limit States: The design of tension members and connections is based on meeting specific limit states, including ultimate limit state (ULS) and serviceability limit state (SLS). The ULS ensures that the structure can safely withstand the maximum expected loads, while the SLS ensures that the structure remains functional and meets serviceability requirements.

d. Design Codes and Standards: Designing tension members and connections involves adhering to relevant design codes and standards. These codes provide guidelines, equations, and safety factors to ensure the structural elements meet specified requirements and performance criteria.

3. Design Methods:

There are various methods and approaches for designing tension members and connections. The selection of the design method depends on factors such as the material used, loading conditions, and structural requirements:

a. Working Stress Design (WSD): The working stress design method involves limiting the stress levels in the material to a fraction of its yield strength. This method is commonly used for materials such as structural steel.

b. Load and Resistance Factor Design (LRFD): The LRFD method, also known as the Load and Resistance Factor Rating (LRFR) method, uses load factors and resistance factors to ensure a sufficient safety margin against failure. It is commonly employed for designing tension members and connections.

c. Strength Design: The strength design method focuses on ensuring that the tension member or connection can safely resist the applied tensile loads without failure. It involves analyzing the ultimate capacity of the member and designing it to exceed the applied loads.

4. Design Process for Tension Members and Connections:

The design process for tension members and connections involves several steps:

a. Determining Design Loads: The first step is to determine the design loads acting on the tension member or connection. These include dead loads, live loads, and any other applicable loads specified by the design criteria.

b. Selecting Appropriate Material: Based on the design loads and structural requirements, an appropriate material is selected. Factors such as strength, corrosion resistance, and cost are considered.

c. Analyzing the Structure: The next step is to analyze the structure to determine the internal forces and moments acting on the tension member or connection. This can be done analytically or by using structural analysis software.

d. Sizing and Designing the Member: Using the determined internal forces, the tension member is sized and designed to resist the applied tensile loads. Design equations, formulas, and code provisions are utilized for this purpose.

e. Detailing and Reinforcement: If the tension member is made of a composite material or requires additional reinforcement, appropriate detailing is carried out based on design requirements. This may include selecting reinforcing bars, determining their spacing, and providing proper anchorage.

f. Connection Design: The design of connections for tension members is crucial for ensuring load transfer and structural integrity. The connection design involves selecting appropriate connection types, determining the connection capacity, and providing proper detailing and reinforcement.

g. Construction Considerations: The final step involves considering construction constraints and detailing requirements for tension members and connections. Proper installation, fabrication, and quality control measures are essential to ensure the performance and safety of the structure.

5. Advanced Design Considerations:

In addition to the basic design considerations, several advanced design aspects can be incorporated for optimizing the performance and efficiency of tension members and connections:

a. Composite Tension Members: Composite tension members, combining different materials such as steel and carbon fiber, can provide enhanced strength, stiffness, and efficiency. Composite action between the different materials is utilized to improve load-carrying capacity.

b. Fatigue Design: In applications where tension members are subjected to cyclic loading, fatigue design considerations are essential. Fatigue life analysis and appropriate detailing are performed to ensure the durability and longevity of the tension members.

c. Sustainability Considerations: Designing for sustainability involves considering factors such as material recyclability, embodied carbon, and life cycle assessment to minimize the environmental impact of tension members and connections.

Conclusion:

The design of tension members and connections is a critical aspect of structural engineering. By considering factors such as load types, material selection, limit states, and design methods, engineers can create safe, efficient, and reliable structures. The design process involves determining the design loads, selecting appropriate materials, analyzing the structure, sizing and designing the tension members, detailing and reinforcing the members and connections, considering construction constraints, and ensuring proper installation. Advanced design considerations, such as composite tension members, fatigue design, and sustainability, can further enhance the performance and efficiency of tension members and connections. By employing sound engineering principles, adhering to design codes and standards, and considering advanced design techniques, engineers can ensure the successful design of tension members and connections in a wide range of structural applications.

11. DESIGN OF REINFORCED CONCRETE STRUCTURES

Reinforced concrete structures are widely used in the construction industry due to their strength, durability, and versatility. From buildings and bridges to dams and foundations, reinforced concrete offers exceptional structural performance. The design of reinforced concrete structures involves a comprehensive process that considers various factors, including load types, material properties, design codes, and construction considerations. This article provides an overview of the design principles and considerations involved in designing reinforced concrete structures, discussing key concepts, design methods, and important considerations.

1. Introduction to Reinforced Concrete Structures:

Reinforced concrete structures consist of concrete elements that are reinforced with steel bars or meshes to enhance their strength and load-carrying capacity. Concrete provides compressive strength, while the steel reinforcement resists tensile forces. The combination of these materials creates a composite structure that can withstand both compression and tension, making reinforced concrete a preferred choice for a wide range of applications.

2. Design Considerations:

The design of reinforced concrete structures involves several considerations to ensure their strength, stability, and overall performance:

a. Load Types: Reinforced concrete structures are designed to resist various types of loads, including dead loads, live loads, wind loads, seismic loads, and temperature effects. Understanding the nature and magnitude of these loads is crucial for proper design.

b. Material Selection: The selection of concrete and reinforcement materials is essential for achieving the desired strength, durability, and performance. Concrete should meet specific strength and durability requirements, while reinforcement bars should possess appropriate strength and ductility properties.

c. Limit States: The design of reinforced concrete structures is based on meeting specific limit states, including the ultimate limit state (ULS) and the serviceability limit state (SLS). The ULS ensures that the structure can safely withstand the maximum expected loads, while the SLS ensures that the structure remains functional and meets serviceability requirements.

d. Design Codes and Standards: Designing reinforced concrete structures involves adhering to relevant design codes and standards. These codes provide guidelines, equations, and safety factors to ensure the structural elements meet specified requirements and performance criteria.

3. Design Methods:

Several design methods and approaches are employed for designing reinforced concrete structures. The selection of the design method depends on factors such as the structural elements, loading conditions, and design requirements:

a. Working Stress Design (WSD): The working stress design method involves limiting the stresses in concrete and reinforcement to a fraction of their respective material strengths. This method considers the service loads and ensures that the stress levels remain within acceptable limits.

b. Load and Resistance Factor Design (LRFD): The LRFD method, also known as the Load and Resistance Factor Rating (LRFR) method, utilizes load factors and resistance factors to provide an increased level of safety against failure. It is commonly employed for designing reinforced concrete structures.

c. Strength Design: The strength design method focuses on ensuring that the reinforced concrete structure can safely resist the applied loads without failure. It involves analyzing the ultimate capacity of the structure and designing it to exceed the applied loads.

4. Design Process for Reinforced Concrete Structures:

The design process for reinforced concrete structures involves several steps:

a. Determining Design Loads: The first step is to determine the design loads acting on the structure. These loads include dead loads, live loads, wind loads, seismic loads, and other applicable loads specified by the design criteria.

b. Selecting Concrete Mix Design: Based on the design loads and structural requirements, an appropriate concrete mix design is selected. Factors such as strength, durability, and workability are considered during the selection process.

c. Analyzing the Structure: The next step is to analyze the structure to determine the internal forces and moments acting on the structural elements. This can be done analytically or by using structural analysis software.

d. Sizing and Designing Structural Elements: Using the determined internal forces, the structural elements such as beams, columns, slabs, and foundations are sized and designed. Design equations, formulas, and code provisions are utilized for this purpose.

e. Detailing and Reinforcement: Detailing and reinforcement design are crucial for reinforced concrete structures. Proper reinforcement detailing, including bar sizes, spacing, and placement, is carried out based on design requirements and construction considerations.

f. Construction Considerations: The final step involves

considering construction constraints and detailing requirements for reinforced concrete structures. Adequate formwork, construction joints, and reinforcement placement details are essential for proper construction and performance.

5. Advanced Design Considerations:

In addition to the basic design considerations, several advanced design aspects can be incorporated for optimizing the performance and efficiency of reinforced concrete structures:

a. Structural Optimization: Structural optimization techniques, such as topology optimization or shape optimization, can be employed to find the most efficient and material-saving configurations for reinforced concrete structures.

b. Prestressed Concrete: Prestressed concrete is a technique that introduces pre-compression to the concrete element, reducing or eliminating tensile stresses and enhancing its load-carrying capacity. Prestressed concrete is commonly used in bridges, parking structures, and high-rise buildings.

c. Sustainable Design: Sustainable design principles are increasingly important in the design of reinforced concrete structures. Considering factors such as material recyclability, embodied carbon, and life cycle assessment can minimize the environmental impact of the structures.

Conclusion:

The design of reinforced concrete structures requires a comprehensive understanding of structural engineering principles, material properties, and construction considerations. By considering factors such as load types, material selection, limit states, and design methods, engineers can create safe, durable, and efficient structures. The design process involves determining the design loads, selecting appropriate materials, analyzing the structure, sizing and designing the structural elements, detailing and reinforcing the elements, considering construction constraints, and ensuring proper installation.

Advanced design considerations, such as structural optimization, prestressed concrete, and sustainable design, can further enhance the performance and efficiency of reinforced concrete structures. By employing sound engineering principles, adhering to design codes and standards, and considering advanced design techniques, engineers can ensure the successful design of reinforced concrete structures in a wide range of construction projects.

12. DESIGN OF STEEL STRUCTURES

Steel structures are widely used in the construction industry due to their strength, versatility, and durability. From buildings and bridges to industrial facilities and offshore structures, steel offers exceptional structural performance and design flexibility. The design of steel structures involves a comprehensive process that considers various factors, including load types, material properties, design codes, and construction considerations. This article provides an overview of the design principles and considerations involved in designing steel structures, discussing key concepts, design methods, and important considerations.

1. Introduction to Steel Structures:

Steel structures are structural systems composed primarily of steel members that carry loads and provide structural stability. Steel possesses excellent properties, such as high strength, ductility, and durability, making it a preferred material for various applications. Steel structures can be fabricated off-site, allowing for efficient construction and easy adaptability.

2. Design Considerations:

The design of steel structures involves several considerations to ensure their strength, stability, and overall performance:

a. Load Types: Steel structures are designed to resist various types of loads, including dead loads, live loads, wind loads, seismic loads, and temperature effects. Understanding the nature and magnitude of these loads is crucial for proper design.

b. Material Selection: The selection of steel grades and sections is essential for achieving the desired strength, ductility, and performance. Steel grades with appropriate strength and toughness properties are chosen based on the design requirements and structural application.

c. Limit States: The design of steel structures is based on meeting specific limit states, including the ultimate limit state (ULS) and the serviceability limit state (SLS). The ULS ensures that the structure can safely withstand the maximum expected loads, while the SLS ensures that the structure remains functional and meets serviceability requirements.

d. Design Codes and Standards: Designing steel structures involves adhering to relevant design codes and standards. These codes provide guidelines, equations, and safety factors to ensure the structural elements meet specified requirements and performance criteria.

3. Design Methods:

Several design methods and approaches are employed for designing steel structures. The selection of the design method depends on factors such as the structural elements, loading conditions, and design requirements:

a. Allowable Stress Design (ASD): The allowable stress design method involves limiting the stresses in steel members to a fraction of their yield strength. This method considers the service loads and ensures that the stress levels remain within acceptable limits.

b. Load and Resistance Factor Design (LRFD): The LRFD method, also known as the Load and Resistance Factor Rating (LRFR) method, utilizes load factors and resistance factors to provide an increased level of safety against failure. It is commonly employed for designing steel structures.

c. Elastic Design: The elastic design method considers the elastic behavior of steel members under loading conditions. It involves analyzing the structure for the applied loads and designing the members to remain within the elastic range.

4. Design Process for Steel Structures:

The design process for steel structures involves several steps:

a. Determining Design Loads: The first step is to determine the design loads acting on the structure. These loads include dead loads, live loads, wind loads, seismic loads, and other applicable loads specified by the design criteria.

b. Selecting Steel Sections: Based on the design loads and structural requirements, appropriate steel sections are selected. Factors such as strength, stiffness, and connection details are considered during the selection process.

c. Analyzing the Structure: The next step is to analyze the structure to determine the internal forces and moments acting on the structural elements. This can be done analytically or by using structural analysis software.

d. Sizing and Designing Structural Elements: Using the determined internal forces, the steel structural elements such as beams, columns, trusses, and connections are sized and designed. Design equations, formulas, and code provisions are utilized for this purpose.

e. Connection Design: The design of connections is crucial for ensuring load transfer and structural integrity in steel structures. Connection types, such as bolted or welded connections, are selected, and appropriate design provisions are followed to ensure their strength and reliability.

f. Detailing and Fabrication: Proper detailing of steel structural elements is essential for fabrication and construction. Detailing includes specifying the member sizes, connection details, and reinforcement requirements.

g. Construction Considerations: The final step involves considering construction constraints and detailing requirements for steel structures. Proper installation, fabrication, and quality control measures are essential to ensure the performance and safety of the structure.

5. Advanced Design Considerations:

In addition to the basic design considerations, several advanced design aspects can be incorporated for optimizing the performance and efficiency of steel structures:

a. Structural Optimization: Structural optimization techniques, such as topology optimization or shape optimization, can be employed to find the most efficient and material-saving configurations for steel structures.

b. Composite Steel Structures: Composite steel structures combine steel and other materials, such as concrete or timber, to enhance structural performance. The combination of materials provides enhanced strength, stiffness, and load-carrying capacity.

c. Fatigue Design: In applications where steel structures are subjected to cyclic loading, fatigue design considerations are essential. Fatigue life analysis and appropriate detailing are performed to ensure the durability and longevity of the steel structure.

d. Sustainability Considerations: Designing for sustainability involves considering factors such as material recyclability, embodied carbon, and life cycle assessment to minimize the environmental impact of steel structures.

Conclusion:

The design of steel structures requires a comprehensive understanding of structural engineering principles, material properties, and construction considerations. By considering factors such as load types, material selection, limit states, and design methods, engineers can create safe, durable, and efficient steel structures. The design process involves determining the design loads, selecting appropriate steel sections, analyzing the structure, sizing and designing the structural elements, detailing and fabricating the elements, considering construction constraints, and ensuring proper installation. Advanced design considerations, such as structural optimization, composite steel structures, fatigue design, and sustainability, can further enhance the performance and efficiency of steel structures. By employing sound engineering principles, adhering to design codes and standards, and considering advanced design techniques, engineers can ensure the successful design of steel structures in a wide range of construction projects.

13. DESIGN OF COMPOSITE STRUCTURES

Composite structures offer unique advantages in the construction industry by combining different materials to create high-performance and efficient structural systems. These structures typically consist of a combination of two or more materials, such as concrete, steel, timber, or fiber-reinforced polymers (FRP), working together to optimize strength, stiffness, durability, and other desirable properties. The design of composite structures involves a comprehensive process that considers various factors, including material selection, load types, design codes, and construction considerations. This article provides an overview of the design principles and considerations involved in designing composite structures, discussing key concepts, design methods, and important considerations.

1.Introduction to Composite Structures:

Composite structures are engineered systems that utilize the advantageous properties of different materials in combination to achieve optimal structural performance. By combining materials with different strengths and characteristics, composite structures can deliver superior strength-to-weight ratios, increased stiffness, and enhanced durability compared to traditional single-material structures.

2. Material Selection:

The selection of materials is a critical consideration in the design of composite structures. The choice of materials depends on the specific requirements of the structure, such as load-carrying capacity, environmental conditions, and desired performance. Common materials used in composite structures include:

a. Concrete: Concrete offers excellent compressive strength and durability, making it suitable for load-bearing elements in composite structures.

b. Steel: Steel provides high tensile strength and stiffness, making it ideal for reinforcement and structural support in composite structures.

c. Timber: Timber offers natural aesthetics and can provide structural support and architectural elements in composite structures.

d. Fiber-Reinforced Polymers (FRP): FRP materials, such as carbon fiber, fiberglass, and aramid fiber, offer high strength-to-weight ratios, corrosion resistance, and design flexibility. They are commonly used as reinforcing elements in composite structures.

3. Design Considerations:

The design of composite structures involves several considerations to ensure their strength, stability, and overall performance:

a. Load Types: Composite structures are designed to resist various types of loads, including dead loads, live loads, wind loads, seismic loads, and temperature effects. Understanding the nature and magnitude of these loads is crucial for proper design.

b. Material Compatibility: Compatibility between different materials in a composite structure is essential to ensure effective load transfer and structural integrity. Material properties, such as thermal expansion coefficients and elastic moduli, must be compatible to minimize differential movement and potential damage.

c. Interface Design: The design of interfaces between different materials in composite structures is crucial for load transfer and overall structural performance. Proper bonding, mechanical connections, and detailing are necessary to ensure efficient load transfer and prevent delamination or separation.

d. Design Codes and Standards: Designing composite structures involves adhering to relevant design codes and standards. These codes provide guidelines, equations, and safety factors to ensure the structural elements meet specified requirements and performance criteria.

4. Design Methods:

Several design methods and approaches are employed for designing composite structures. The selection of the design method depends on factors such as the structural elements, loading conditions, and design requirements:

a. Strength Design: Strength design involves designing composite structures based on the strength capacity of the materials and the applied loads. This method ensures that the structure can safely carry the expected loads without failure.

b. Allowable Stress Design (ASD): The allowable stress design method involves limiting the stresses in the materials to a fraction of their allowable strengths. This method considers the service loads and ensures that the stress levels remain within acceptable limits.

c. Performance-Based Design: Performance-based design aims to optimize the performance of composite structures by considering specific performance objectives, such as strength, deflection, or durability. This method involves conducting advanced analyses and performance evaluations to achieve desired outcomes.

5. Design Process for Composite Structures:

The design process for composite structures involves several steps:

a. Determining Design Loads: The first step is to determine the design loads acting on the structure. These loads include dead loads, live loads, wind loads, seismic loads, and other applicable loads specified by the design criteria.

b. Material Selection: Based on the design loads and structural requirements, appropriate materials are selected for different components of the composite structure. Considerations such as strength, stiffness, durability, and compatibility are taken into account.

c. Analyzing the Structure: The next step is to analyze the composite structure to determine the internal forces and moments acting on the structural elements. This can be done analytically or by using structural analysis software.

d. Sizing and Designing Structural Elements: Using the determined internal forces, the structural elements, such as beams, columns, slabs, and connections, are sized and designed. Design equations, formulas, and code provisions are utilized for this purpose.

e. Connection Design: The design of connections in composite structures is crucial for ensuring load transfer and structural integrity. Connection types, such as mechanical fasteners or adhesives, are selected, and appropriate design provisions are followed to ensure their strength and reliability.

f. Detailing and Fabrication: Proper detailing of composite structural elements is essential for fabrication and construction. Detailing includes specifying the component sizes, connection details, and reinforcement requirements.

g. Construction Considerations: The final step involves considering construction constraints and detailing requirements for composite structures. Proper installation, fabrication, and quality control measures are essential to ensure the performance and safety of the structure.

6. Advanced Design Considerations:

In addition to the basic design considerations, several advanced design aspects can be incorporated for optimizing the performance and efficiency of composite structures:

a. Structural Optimization: Structural optimization techniques, such as topology optimization or shape optimization, can be employed to find the most efficient and material-saving configurations for composite structures.

b. Sustainable Design: Sustainable design principles are increasingly important in the design of composite structures. Considering factors such as material recyclability, embodied carbon, and life cycle assessment can minimize the environmental impact of the structures.

c. Durability Design: Composite structures may be exposed to various environmental conditions, such as moisture, chemicals, or temperature variations. Designing for durability involves selecting materials and protective measures to ensure long-term performance and resistance to degradation.

Conclusion:

The design of composite structures requires a comprehensive understanding of structural engineering principles, material properties, and construction considerations. By considering factors such as material selection, load types, design codes, and construction constraints, engineers can create innovative and efficient composite structures. The design process involves determining the design loads, selecting appropriate materials, analyzing the structure, sizing and designing the structural elements, detailing and fabricating the elements, considering construction constraints, and ensuring proper installation. Advanced design considerations, such as structural optimization, sustainable design, and durability design, can further enhance the performance and efficiency of composite structures. By employing sound engineering principles, adhering to design codes and standards, and considering advanced design techniques, engineers can ensure the successful design of composite structures in a wide range of construction projects.

14. STRUCTURAL DYNAMICS AND EARTHQUAKE ANALYSIS

Structural dynamics and earthquake analysis are critical components of structural engineering that focus on understanding the behavior of structures under dynamic loads, particularly seismic events. Earthquakes pose significant challenges to the safety and stability of structures, making it essential to design and analyze them to withstand the forces generated by such events. This article provides an overview of structural dynamics and earthquake analysis, discussing key concepts, analytical methods, and design considerations.

1. Introduction to Structural Dynamics:

Structural dynamics deals with the study of structures' response to dynamic loads, including vibrations, impact forces, and seismic events. It considers the behavior of structures under time-varying loads, differentiating it from static analysis, which focuses on equilibrium under constant loads. Structural dynamics plays a vital role in designing structures that can withstand various dynamic loads, including earthquakes, wind, and machinery-induced vibrations.

2. Earthquake Characteristics:

Earthquakes result from the sudden release of energy in the Earth's crust, causing ground shaking and propagating seismic waves. Understanding the characteristics of earthquakes is crucial for seismic analysis and design:

a. Magnitude: Earthquakes are measured on the Richter scale, which quantifies their energy release. Magnitude determines the severity of an earthquake, ranging from minor tremors to major destructive events.

b. Frequency Content: Earthquakes generate waves with different frequencies, including high-frequency waves that cause rapid ground motion and low-frequency waves that induce long-period oscillations in structures.

c. Ground Motion: Ground motion refers to the movement of the Earth's surface during an earthquake. It is characterized by parameters such as amplitude, duration, and frequency content.

3. Seismic Analysis Methods:

Seismic analysis involves evaluating a structure's response to ground motion and ensuring its stability and safety. Various analytical methods are employed for seismic analysis:

a. Equivalent Static Analysis: Equivalent static analysis approximates the dynamic effects of an earthquake by applying an equivalent static load. This method is commonly used for low-to-moderate seismicity regions and simpler structures.

b. Response Spectrum Analysis: Response spectrum analysis utilizes the response spectrum, which represents the structure's response to a range of ground motion frequencies. It provides a more accurate assessment of a structure's behavior under seismic loads.

c. Time History Analysis: Time history analysis simulates the actual ground motion recorded during an earthquake. It considers the complete time history of ground motion, allowing for a detailed assessment of the structure's dynamic response.

4. Design Considerations:

Designing structures to withstand earthquakes requires careful consideration of various factors:

a. Building Codes and Regulations: Design codes and regulations provide guidelines for seismic design and ensure structures meet minimum safety requirements. These codes specify factors such as seismic design categories, load combinations, and design methodologies.

b. Seismic Hazard Assessment: Seismic hazard assessment involves evaluating the level of seismicity in a given region and estimating the potential ground motion. This assessment helps determine the appropriate design parameters and ground motion inputs for seismic analysis.

c. Structural Dynamics: Understanding the dynamic behavior of structures is essential for seismic design. Structural dynamics considerations include natural frequencies, mode shapes, damping, and response characteristics.

d. Ductility and Energy Dissipation: Structures designed for seismic loads should exhibit sufficient ductility to absorb and dissipate energy during an earthquake. Ductile materials and detailing techniques are employed to enhance the structure's ability to withstand seismic forces.

e. Base Isolation and Damping: Base isolation systems and damping devices can be incorporated into the design to reduce the transmission of seismic forces to the superstructure. These systems help improve the structure's response and mitigate damage.

f. Soil-Structure Interaction: The interaction between the soil and the structure plays a significant role in seismic response. Soil conditions, including soil type, site amplification effects, and liquefaction potential, should be considered in the design process.

5. Seismic Design Philosophy:

The seismic design philosophy aims to ensure that structures can withstand the expected seismic forces and limit damage during an earthquake. Key principles of seismic design include:

a. Life Safety: The primary objective of seismic design is to protect human life by ensuring that structures remain stable and do not collapse during an earthquake.

b. Structural Integrity: Seismic design focuses on maintaining the structural integrity of the building, allowing it to withstand the expected seismic forces without significant damage.

c. Functional Continuity: Seismic design aims to maintain the functionality of critical facilities during and after an earthquake, enabling essential services to continue.

6. Advancements in Earthquake Engineering:

Advancements in earthquake engineering have led to improved seismic design practices and mitigation strategies:

a. Performance-Based Design: Performance-based design focuses on achieving desired performance objectives, such as limiting damage or maintaining functionality, rather than solely meeting prescriptive code requirements.

b. Seismic Retrofitting: Retrofitting involves strengthening existing structures to enhance their seismic resistance. Techniques include adding shear walls, installing bracing systems, or applying external damping devices.

c. Seismic Isolation: Seismic isolation involves incorporating isolation bearings or flexible elements to decouple the superstructure from the ground, reducing the transmission of seismic forces.

d. Advanced Analytical Techniques: Advanced numerical methods, such as finite element analysis and nonlinear dynamic analysis, allow for more accurate predictions of a structure's behavior under seismic loads.

Conclusion:

Structural dynamics and earthquake analysis are critical aspects of structural engineering, particularly in regions prone to seismic activity. Understanding the behavior of structures under dynamic loads and designing them to withstand earthquakes is essential for ensuring the safety and integrity of buildings and infrastructure. Seismic analysis methods, such as equivalent static analysis, response spectrum analysis, and time history analysis, help assess the structural response to ground motion. Design considerations include adherence to building codes and regulations, seismic hazard assessment, structural dynamics, ductility, energy dissipation, soil-structure interaction, and the incorporation of base isolation or damping systems. With advancements in earthquake engineering and the adoption of performance-based design approaches, structures can be designed and retrofitted to better withstand seismic forces, reducing the risk to life and property during earthquakes. By considering these factors and employing sound engineering principles, structural engineers can contribute to the development of resilient and earthquake-resistant structures that safeguard communities in seismic-prone areas.

15. INTRODUCTION TO STRUCTURAL DESIGN CODES AND STANDARDS

Structural design codes and standards play a fundamental role in the field of structural engineering. They provide guidelines, criteria, and specifications for the design, construction, and maintenance of structures, ensuring their safety, reliability, and compliance with regulatory requirements. This article provides an introduction to structural design codes and standards, discussing their significance, development, and key components.

Importance of Structural Design Codes and Standards:

Structural design codes and standards are essential for several reasons:

a. Safety: The primary objective of design codes and standards is to ensure the safety of structures and protect the lives of occupants and users. They establish minimum requirements for structural performance and load resistance to prevent structural failure and collapse.

b. Reliability: Codes and standards provide a framework for designing structures that can withstand anticipated loads and environmental conditions over their expected service life. They ensure the reliability and performance of structures, minimizing the risk of unexpected failures.

c. Uniformity: Design codes and standards promote uniformity and consistency in the design and construction of structures. They provide a common language and set of guidelines that engineers, architects, contractors, and regulatory authorities can follow, facilitating effective communication and understanding.

d. Regulatory Compliance: Codes and standards are often mandated by regulatory bodies and government agencies. Compliance with these requirements is necessary to obtain building permits and approvals, ensuring that structures meet the necessary safety and quality standards.

Development of Structural Design Codes and Standards:

Structural design codes and standards are developed through a rigorous and collaborative process involving industry experts, researchers, professional organizations, and regulatory bodies. The development process typically includes the following stages:

a. Research and Technical Basis: The development of codes and standards begins with research and the establishment of a technical basis. Experts review existing knowledge, conduct experiments, and perform analytical studies to determine the best practices and design methodologies.

b. Committee Formation: Committees or panels consisting of professionals from academia, industry, and regulatory bodies are formed to oversee the development process. These committees review the technical information, discuss the proposed changes or updates, and draft the code provisions.

c. Public Input and Review: The draft codes and standards are made available for public input and review. This allows stakeholders, including engineers, architects, contractors, and the general public, to provide feedback, suggestions, and comments on the proposed changes.

d. Consensus and Approval: The committees consider the public input and make revisions to the draft codes and standards. Consensus is reached through discussions and deliberations. Once the final version is approved, it is published and made available for implementation.

e. Periodic Updates: Codes and standards are periodically updated to incorporate advancements in technology, research findings, and lessons learned from past experiences. This ensures that the codes remain relevant and up-to-date with the latest industry practices.

Components of Structural Design Codes and Standards:

Structural design codes and standards typically consist of several components that provide comprehensive guidance for the design process:

a. General Requirements: The codes and standards include general requirements that outline the scope, objectives, and application of the document. They define the responsibilities of the design professionals, construction practices, and quality control requirements.

b. Design Loads: Design codes specify the types of loads that structures are expected to resist. These include dead loads (e.g., self-weight of the structure), live loads (e.g., occupancy loads), wind loads, seismic loads, snow loads, and others. The codes provide specific load combinations and load factors for different design scenarios.

c. Design Criteria and Methods: Codes and standards establish design criteria and methods for analyzing and designing structural elements. This includes provisions for structural stability, strength, serviceability, and durability. The design methodologies may vary based on the material types, structural systems, and load conditions.

d. Material Specifications: Codes provide specifications for the selection and use of construction materials, including concrete, steel, timber, masonry, and composites. They define the material properties, testing methods, and quality control measures to ensure the proper use of materials in construction.

e. Construction and Quality Control: Design codes include provisions for construction practices, inspection procedures, and quality control measures. They outline the requirements for proper execution of design details, connections, and construction techniques to ensure the integrity and safety of the structure.

f. Performance and Design Verification: Some codes and standards incorporate performance-based design approaches, allowing engineers to assess the structure's performance based on specific performance criteria. Verification procedures, such as structural testing, numerical simulations, or analytical methods, may be provided to ensure the structure meets the desired performance objectives.

International and Regional Codes:

Structural design codes and standards can vary between countries and regions due to variations in climate, construction practices, and regulatory frameworks. Some international organizations, such as the International Building Code (IBC), International Organization for Standardization (ISO), and Eurocode, provide guidelines that are widely adopted and recognized globally. These international codes serve as references for countries in developing their national or regional codes.

Conclusion:

Structural design codes and standards are vital tools in the field of structural engineering, ensuring the safety, reliability, and uniformity of structures. They provide guidelines for the design, construction, and maintenance of structures, incorporating requirements for load resistance, structural analysis, material specifications, construction practices, and quality control. The development of these codes involves research, consensus-building, and periodic updates to incorporate advancements in technology and industry practices. Compliance with design codes and standards is crucial for obtaining building permits and ensuring regulatory compliance. By adhering to these guidelines, structural engineers can create structures that meet the necessary safety and quality standards, contributing to the resilience and sustainability of the built environment.

Printed in Great Britain
by Amazon